Stems and Leaves
poems inspired by programming

Rorie Newman

Layout by Rorie Newman
The font is Garamond, 12 point

CONTENTS

INTRODUCTION

As a beginner in Computer Science and Programming classes, I struggled academically as well as struggled to accept that I was doing poorly in a subject that I enjoyed. I couldn't wrap my brain around theoretical math or data structures. However, at the end of one of my courses, we were given the opportunity to create a project revolving around a topic, structure, or language we had not covered in the course. It was a project I was initially dreading: *If I can't understand these topics when I'm being taught by a professor, how am I supposed to learn one on my own?* I decided to try and produce a poem on splay trees, a data structure. As I began to do research for my piece, I realized that I was actually understanding what I was learning. A light bulb lit up in my mind. I discovered that I needed to indulge my more art-oriented brain using metaphors to understand the very analytical concepts and to develop an emotional connection to the subject matter. Writing a poem about a splay tree neglected by its romantic partner and trapped in a one-sided emotionally abusive relationship, made me feel more confident than I've felt in any STEM assignment I've tackled since coming to college.

This project inspired the premise of this chapbook. Through the generosity of the Kirkland Endowment Advisory Committee and under the guidance of my professor Naomi Guttman, I was able to create this miniature collection. I structured my pieces to follow my emotional journey of learning to program from insecurity to anger and frustration to love and contentment. I pair traditionally formed love poems with the emotions expressed in my free-form metaphor-based programming poems because I wanted my poems to be accessible to someone with little to no programming experience, but also to serve as fun puzzles for those who may be more experienced in the realm of computers.

This project really allowed me to dive into specific data structures, algorithms, languages that I was interested in and conceptualize them in new ways that made sense to me. As a result, I was able to grow my understanding of these concepts and to grow my poetic skills at the same time. And in getting to combine two academic areas I love, I was able to really enjoy what I was doing. I put my whole heart, and brain, into this collection and I hope it can resonate with other people struggling in the early beginnings of programming, but also with people who may be struggling in other academic areas, learning how to balance the love of the subject with the difficulty faced in exploring it.

TERMINAL

takes in the input when running a program and alerts the programmer of errors in the code

the end
is not dark with a quick bright light
gray and quick
like a computer shutting down in an 80s movie
instead
it is this slow stuffy press of time that is warm
slightly too warm
because the bus terminal   bus stop   manages
to be crowded
even though its outside, this stuffy press is slow
arduous, long
even though there's a schedule
the bus is late

even though we are at the terminal
stops
the trip is only beginning
the end
is filled with waiting
uncomfortable
a ghost drifting from one terminal to the next
waiting
a wait whose only relief is a bus where scratchy seats and flashing
window vignettes
remind us of how far we still are from home, from the
the next stop

even the Romans, wise as they were didn't
know

if the terminal was the end or the beginning
Terminus
sets his boundaries, but do the boundaries mark the end of
what's known
or do they mark the beginning of
what's to come
and if they serve as boundaries for our travels
then why
does each pastoral scene, each bus stop sign send
unanswerable warnings
to our minds, our nerve endings receiving a never ending stream
of errors
"aren't we home yet?"

maybe
the Romans did know
maybe it's both
maybe it's   everything
it's the end of everything that we know
will happen
it's the ferrying, the waiting of bus stop passengers
it's the ghosts
it's each warning sign
making the end, making home
so much more
it's the next stop
it's always the next stop
it's the boundary and it's
the beginning

DRESSED IN GREEN

Of everything that is the color Green–
the trees, the moss, the stones, and the mild seas,
the apples dancing: a granny Green scene–
You, my love, are the Green I want to seize.
You, the shoreline, where the blue laps the land
You, where the tree trunks bend–I want to wrap
them around my waist, held safe in Your hand–
where seeds flutter, inventing their own map.
And You, Your smile, Your happy, is the sun
his rays twisting, dancing golden and strong
through Green spears, I'm afraid to touch, to want
of someone else's briefly-fresh mowed lawn.
And now I finally understand why
Green is seen as envy. I vie to hoard.

SPLAYED OPEN

the splay tree data structure, which functions similarly to a binary search tree, but with the most notable difference in functioning being the principle of locality, where the last piece of data accessed becomes the new root

I change
Myself   for you
Molding   my body for you
Zig   pulling my limbs
Zig Zag   turning myself upside down and backwards
Zig Zig   pulling and pulling
And   turning and turning
So you   can access
The parts of me   you claim you love
And not   my faults
          You choose to ignore

You choose to ignore
Me   in favor
Of others   of their consistency
          of their beauty
Like trees   of red and black
Blooming   in colors I don't don
Balanced   in ways my legs won't support
Alterable   in a way I still resist
          as much as I can
But you   you like how easy it is
When   You search my body

You search my body
For the   dips and joints

And swatches   of skin that appeal
        to you
You delete
The   dips and joints
And swatches   of skin that do not suit your needs
And you insert
New   dips and joints
And swatches   of skin that you like better
Than what   I already am
In doing this   you take your time
        Sometimes
In doing this   you sift through parts of me
        Sometimes
        until you find what you need
In doing this   you rush Sometimes
        Knowing
What you desire   and taking it quickly
Yet never   as quickly as you'd like
As quickly   as you can get it
From those   who at their worst
Are   better than me
At my   worst so
Why   Would you ever pick me

Would you ever pick me
I don't know   if I would even pick me
Because I create words   that don't exist
That don't come   from your mouth
Or   your fingers
As I analyze   each letter of everything
        You say
Playing on   repeat in my mind
Trying   to predict what may come next

But I   hoard our memories like pretty stones
Your favorite   at the top of my cache
Ready   and waiting
For you   to want to
      look
And   Remember

Remember
I   change
Myself   for you
Molding   my body for you
Zig   pulling my limbs
Zig Zag   turning myself upside down and backwards
Zig Zig   pulling and pulling
And   turning and turning
So you   can access
The   parts
      of Me
You claim   You love

A ONE SINNER CONGREGATION

It never occurred to me that I'd be
vying against someone who doesn't exist.
That I'd overthink every kiss because
I haven't done enough to earn your heart's key.
That although I am your humble devotee
you deserve so much more than my list
of flaws that remain, draining my—your—glee.
        I'm a dark cloud over us, a heavy mist.

But you look at me like I'm the sun
with soft sherbet haloing my frame;
like I'm lightning – an exciting phenom;
someone to cherish, not to blame.
So for a second, I don't feel like I'm a con,
        and let myself dissolve in your warm flame.

!

the process of programming factorials, where each number in the sequence equals the product of all the numbers before it

fact 1: i love you

fact 2: it was impossible not to love you
        not to love
        your straightforward beauty
        your simplistic ease
        the way you tell me
        that you're mine
        that your heart
        is a puzzle
        that I solved
        on my first try

fact 3: it was impossible not to love you if i just ignored
        the first few times i mixed up
        whether i was supposed to add
        helpful advice when you're struggling
        when your problems themselves
        add up
        or if i was supposed to multiply
        the distractions, to divert your mind
        from its negative focus
        i'm supposed to ignore it
        so why did you
                why do you
        keep
        reminding
        me

keep bringing up my errors
and shooting warning glances
as my composure fractions into dividends

fact 4: it was impossible not to love your straightforward lines if i
just ignored
    that they were masking lies
    hiding what lies
    within your complicated mind
    i'm Sisyphus straining against the rock
    it's heavy
    it keeps rolling back and squishing my toes
    my shoes fill with mud and dirt
    and i
    don't know whether to focus
    on the rock
    or my toes
    or my shoes
    or
            you

fact 5: i hate you

fact 6: i hate that i just can't ignore
    the screaming warning signs
    ignore
    your screaming
    no matter how many times
    i want to
    i can't ignore your words
    soft
    gentle
    saying you're almost there

followed by
more screaming

fact 7: i hate that i just can't ignore the sometimes
   the sometimes it works out
   evens out
   we answer
   one
   or two
   maybe even three
   of our problems
   your problems
   moving the rock a little further
   bandaging my toes
   and washing my sneakers of mud

fact 8: i will always love you

   or i
   always lie
   because i won't really
   always
   love you
   but i'll always think
   of you
   always have your unfinished sudoku puzzle
   resting behind my eyes
   swapping numbers while
   I'm trying to sleep
   never finding the right one

   every ex, every solver, every lover

must have placed every number where it belongs
found the right spot for every nine
found the solution
found you

because as easy as you made it seem
       with everyone else

fact 9: maybe your heart
      was never my puzzle to solve

    at least
    not
    for right now

NOT A LOVE POEM

I will not use the word "love" in my love poem and I will not call him the sun or the stars or the flowers that bloom or the thousands of other clichés that dance in my mind when I see him. Maybe I could call him the ocean, peaceful, calming: a gentle hum, but water is as washed up as roses or the moon. What if I compared him to trees, standing steady against the world, tall and strong, but leaves all a rustle by the wind, by me. Yet, this too is worn from weathering. I am not implying that he is boring or that he evokes yawning clichés. I am simply overcome with mountains of feelings, repetitive images layered on top of me, stones and dirt demanding my focus. Feelings which make my mind grow fuzzy: I can only think of him in cliches. I want to call him the sun, to say his bright smile causes my heart to dum-dee-dum. He makes me dumb, in the most frustrating ways, in ways that cause poetry to gush from my mind, my throat. But he deserves something new, the newest syntax, the newest code, the newest technique, concept, and tool. And regardless of how much I want to, or how much strength it takes to create this new: I will not use the word "love" in my love poem.

DEAR RECURSION

a programming technique where a function is called within itself, and builds on the previous call until a given condition is met

you are a lawyer bringing up defenses
in the middle of spats on the same issues
adding pieces and new information
to build off of–only to bring that argument
up again

do you even remember the original point you were trying to make?

you are a choose your own adventure story
forcing your protagonist to take detours
through paths they've already traveled,
but armed with new words, new characters,
new plot devices; and they walk the path
of words again

do you even have strength left in your feet to make another go?

you are a car looping around a block
and like a car you need, you need to be
picking up new passengers, replacing
old drivers and looping until
you hit a stop sign, until there are no
more passengers to pick up, until you
hit a red light–and the driver
waits
to turn left instead of right

are you even able to stop on your own? if you could, would you?

you need to be interrupted
by a defendant, a judge
you need your protagonist to reach
the end of the path, the chapter, the book
you are someone who needs
a stop sign
just as much as the passengers

                                        need you

IF ONLY I COULD FORGET
YOU[R TOUCH]

Oh to carry a poppy like Hypnos.
I would spend my days in my lonely cave
sleeping away my yearnings to hold you close,
to taste your sweet skin once again and relieve these cravings.

I would spend my days in my lonely cave
with dismal thoughts raining like pollen, so morose–
to taste your sweet skin once again and relieve these cravings–
the only heat in this darkness, my senses overthrown.

With dismal thoughts raining like pollen, so morose,
I bathe in Lethe to clear my skin of memories you engraved.
The only heat in this darkness, my senses overthrown
by the river's arms, its lulls, its gentle waves.

I bathe in Lethe to clear my skin of memories you engraved,
sleeping away my yearnings to hold you close,
        Lethe–your arms, your lulls, your gentle waves.
Oh to carry a poppy like Hypnos

LOVE LANGUAGE
time complexity, how efficiently or quickly a function runs, of common algorithms in Big O

$O(n^2)$ - words of affirmation
love notes fall from my mouth showering you
in painless paper cuts, like leaves from an autumn
tree raining down around, the reds, and oranges,
and browns forming nests at your feet of nested
sweet nothings that mean everything. for each action
you take, each thought you share, each determined step
you take—worried about stumbling, the way a bird, a
hatchling, worries about taking flight, but stepping, flying,
anyways: pamphlets of praise spill slowly. slowly not
because i need the time to form my fanfared phrases,
but because there are simply too many to be quick. so
steadily, i twitter and chirp until my words become
an extra breeze keeping your wings—keeping you—aloft
and soaring, through the trees.

$O(nlgn)$ - gift giving
i search stores and buy craft supplies to form
the perfect gift, narrowing my selection criteria
inch by inch, until i'm left with one, perfect, object
for you. because for you what else would i give—
what else do you deserve—besides perfect. and i
spend just as long—longer—to bestow it upon you
and i will repeat the process as long as it takes
until you're buried in a pile of glitter glue and cut-out
hearts, a pile of legos to build your favorite animal,
a pile of 'i saw this and it made me think of you',
a pile of love and until you feel worthy. because you are.

O(n) - physical touch
i hold your hand, palms pressing together as if in prayer.
i kiss every inch of your thighs, my mouth moving and
mumbling your name as if it belonged to a saint,
until you forget it belongs to you, utterly lost
among the heat of the path i'm tracing. i spend time with
every inch of your body, the same amount of time,
the same amount of reverence. because how could i not
cherish the forearms, the stomach, the kind mind
of the person i love; every part of the incredible whole.

O(lgn) - quality time
i search you out, casting my line, and removing
commitments like pond grass. i save my energy for
when my line tugs, for when it's time to reel. or
even when the fish refuse to bite, and we tool around
in fruitless circles. which is never a chore. not when
it's you. and we sail together, taking our time.
together.

O(1) - acts of service
the cost to relieve your daily burdens, to make you
smile: the penny fractions advertised at gas stations.
an amount so little it refuses to add up. and even if it did?
it'd be worth it.

because as the sun nears its journey's end, the moon
beginning to glitter among the darkening sky, the
nightingales circle and sing, and our lines have
long grown lonely, a little prayer leaves my mouth
and i toss a penny into the river like a fountain
wishing for nothing more than what we are

LIGHTBULB LOVE

My love was not love at first sight nor was
it even love at fifth or sixth sight. It
blossomed from a lightbulb moment; because
you remembered some little thing I told
you in passing. That bright yellow light guided
my eyes and I could see that we were never
apart. That I never wanted to be. I could see eyes
that softened like butter, just for me. I saw
hands that I wanted to hold. I could feel that our
knotted cords were beating to the tempo of
'what ifs.' And I felt they were starting to
untangle. Until only a single string, a
'what if it's perfect' remained. And with our two
hearts linked, I promised to never let you go.

RUBY IN A BOTTLE

the ruby programming language which uniquely integrates functional (runs in order of functions called), procedural (runs from top to bottom), and object-oriented programming (uses a series of classes and objects)

a bottle, a message bottle
    bobs up, bobs down
    riding the waves
    glass fogging from the sun's heat
    from the cool water
    hiding the object
    resting inside
    what rocks inside
    back and forth in time
    following the motion
    of its container
    of its superclass

a ruby red ring
    a gold band
    with a ruby set in prongs
    a ruby not a pearl
    it's far too structured to be a pearl
    smooth facets functioning
    as a prism
    reflecting the light
    streaming through the bottle
    across its paper companion
    across its subclass
    allowing it to be read
    through a ruby lens

a poem, a contrapuntal
    a set of three
    lined up in pretty rows
    language far simpler
    than the poetic musings of Romantics
    or Elizabethans
    to be read top to bottom or
    to be read across
    messages dancing between functions
    letting the words carry themes
    and data between stanzas
    different poems in the set
    but regardless
    forming meaning from magic
    an interpreters dream

a bottle, a message bottle
a ruby red ring
a poem, a contrapuntal
rocking
    with the tumultuous waves
    bob up
    roll forth
    bob down
    roll back
rocking

MUSHROOM MESSAGES

messages in the smalltalk programming language,
which consist of three types, all of which have a
receiver and a message selector telling the receiver
what to do

in the forest,
a pine tree grove,
mushrooms grow
spiral staircases
along bark and boughs.
here the trees talk
back and forth
in mushroom messages.

the network of spores
guides signals sent
between trees.
signals detailing
the pine tree receiver,
and the action
needed to be taken.
the trees send
requests for relief
and words of warning
like "more nitrogen"
or "release hormones;
watch for toxins"
in toadstool telegrams.

the trees' mutualistic
friends decipher
these messages

and send them exactly
where they need to go.
knowing which trees
aid and which
trees don't even if
the sender doesn't.
stamping "error"
like "return to sender"
they work like
porcini postmen.

in the pine tree grove,
it's not the straight-backed
trees in their crowns
of leaves
that command the
wilderness to grow,
but the mushrooms
weaving tree to tree
connecting their roots
together under soil rich
with knowledge and meaning
allowing them,
the trees, to function.
it's the mushrooms
that let them grow
the mushrooms and their
messages, their humble
fungus forecasts.

PYTHON: A SNAKE'S MYTH

a beginner friendly programming language with easy-to-understand syntax and versatile uses which make it a popular choice for experienced programmers as well

There is
a python
   a snake
who slithers
slowly   her
scales scratching
through the
dancing pictures
inside softly
sleeping heads

she visits
bringing easy
dreams   understood
not as
omens but
as lessons
   a first
taste of
how to
function in
the wide
wide world
   how to
loop through
cities   through
scenarios unscathed

she provides
maps   laying
the groundwork
for journeys
both near
and far
    she is
a rite
of passage
    visiting virgin
travelers and
visiting experts
in wanderlust
    a comfortable
safety net
    a snake
who slithers

slowly   her
scales scratch
through lines
of streets
    through mountain
paths   across
coastal shores
    not always
quite remembering
where she
came from
but always
knowing where
to go

she does
not shed
her skin
    no   instead
she builds
her own
exoskeleton of
layered knowledge
    of hardened
navigation   getting
her to
where she
plans to
go   where
she needs
to go

and she
is beautiful
    her presence
dancing through
the unconsciousness
of those
who welcome
her return
her ease
her simplicity
her vast
experience   her
knowledge and
they seldom
let her
slip away

There is
a python
    a snake
who slithers
slowly   her
scales scratching
minds both
old and
new

LEAP FROG

the bubble sort algorithm, which is a slow sorting algorithm that functions by running through a set and comparing two adjacent pieces of data, conditionally switching their places to put them in order

the pond is not blue, nor green, but settles somewhere in between // at least I think it settles in between // the color I guess doesn't matter all that much // no one cares what color the serving platter is // only that what's on it is delivered quickly

the pond is a platter // a fancy silver platter, mirroring the bushes and reeds beside lily pads // lily pads which glide back and forth on their anchors, backs carrying blooming white flowers // and on each pad, beside each flower is a frog // green like only a frog can be, that muted color dancing in hues of what should have been brown // speckled in the green of leaves that hang on black cherry trees

and the frogs they dance in their lily pad line // deep croaks from their chests // from their bellies // from their hearts // from where they leap over one another // but one at a time, taking turns // they leap to their own tune, taking their time // a tune that is slow // too slow for the jumps they are taking // the first jump, the first frog, then next and next // and over // and over // and over again

until Nature's patterns form art out of coincidence // is it really coincidence if it was planned // no, but it feels that way regardless as each frog is arranged from small to large

how funny Ms. Nature is // to create this uncanny, infallible world // with everything placed right where it should

COMING TO

There are no butterflies when I see you
but with you, after my chest has been squeezed
I can finally breathe; my lungs come to

I'm sorry my heart doesn't flutter anew
no wings held by an electrified breeze
There are no butterflies when I see you

My heart, though, is yours controlled by voodoo
in a jar marked with your name, with expertise
I can finally breathe; my lungs come to

You believe my nerves are still: a statue
tall, proud, and cold as naked winter trees
There are no butterflies when I see you

However, my heart does beat, my blood brews
because your smile, all dimpled and wide, gives me a reprieve
I can finally breathe; my lungs come to

You're a warm road, a chalked street avenue
and the five pm sound of jangling keys
There are no butterflies when I see you
But I can finally breathe; my lungs come to

www.ingramcontent.com/pod-product-compliance
Lightning Source LLC
Chambersburg PA
CBHW070454130626
46553CB00006B/2405